Housing
84

感谢你的音乐

Thank You for the Music

Gunter Pauli

[比] 冈特·鲍利　著

[哥伦] 凯瑟琳娜·巴赫　绘

唐继荣　译

上海远东出版社

丛书编委会

主　任：田成川

副主任：何家振　闫世东　林　玉

委　员：李原原　翟致信　靳增江　史国鹏　梁雅丽

　　　　任泽林　陈　卫　薛　梅　王　岢　郑循如

　　　　彭　勇　王梦雨

特别感谢以下热心人士对童书工作的支持：

匡志强　宋小华　解　东　厉　云　李　婧　庞英元

李　阳　刘　丹　冯家宝　熊彩虹　罗淑怡　旷　婉

杨　荣　刘学振　何圣霖　廖清州　谭燕宁　王　征

李　杰　韦小宏　欧　亮　陈强林　陈　果　寿颖慧

罗　佳　傅　俊　白永喆　戴　虹

目录

Contents

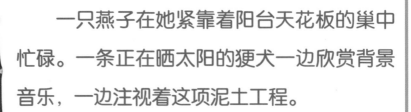

一只燕子在她紧靠着阳台天花板的巢中忙碌。一条正在晒太阳的猂犬一边欣赏背景音乐，一边注视着这项泥土工程。

"你筑巢这么忙，可能没有任何空闲时间为我唱歌了。"猂犬说。

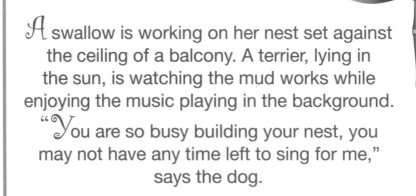

A swallow is working on her nest set against the ceiling of a balcony. A terrier, lying in the sun, is watching the mud works while enjoying the music playing in the background.

"You are so busy building your nest, you may not have any time left to sing for me," says the dog.

你可能没有任何空闲时间为我唱歌了

You may not have any time left to sing for me

我们鸟儿一直会找时间来唱歌

We birds always find time to sing

"哦，不会的！我们鸟儿一直会找时间来唱我们最优美的歌曲。"燕子回答道。"但现在我必须建一个安全的场所来下蛋。"

"那么，你什么时候有空，能让我再次欣赏你甜美的歌声？"㹴犬问。

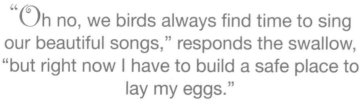

"Oh no, we birds always find time to sing our beautiful songs," responds the swallow, "but right now I have to build a safe place to lay my eggs."

"So when will you have time to let me enjoy the sweetness of your voice again?" asks the terrier.

"听着！鸟儿并不会为别人唱歌，我们只在自己想要唱歌的时候才唱。当太阳升起时，当月光激发我们的情感时，或当鲜花盛开且食物丰盛时——任何我们想分享生活中的快乐的时候，我们就唱歌。"

"你和伙伴们的歌声让我非常陶醉，你们已经启发了世界上最美好的作品。全人类历史上最受尊敬的作曲家莫扎特，就在他的一部钢琴协奏曲中加入了紫翅椋鸟的歌声。"

"Look here, birds do not organise concerts for others, we sing when we feel like it. When the sun rises, when the moon inspires us, or when the flowers are out and food is abundant. Whenever we like to share the joys of life, we just sing."

"I am so enchanted by your songs, and those of your fellows. You have inspired the best in the world. Like Wolfgang Amadeus Mozart, the most admired composer of all times, who included a starling's song in one of his piano concertos."

Wolfgang Amadeus Mozart

... the great mathematician!

"是啊，那人非常著名。但那是紫翅椋鸟，不是燕子。即便这样，我们也为自己的同类激发了伟大数学家的灵感而骄傲。"

"你是指音乐家吧！"

"Oh yes, that one is famous. But that was a starling, not a swallow. Still we are very proud that one of our brothers inspired the great mathematician."

"You mean musician!"

"莫扎特是一位伟大的数学家。他的作品节奏影响人们的脑电波，帮助孩子们学习。当他们在欣赏古典音乐时，几何和代数会学得更好。"

"真不可思议！要知道，长久以来，人类认为飞行只需要羽毛和翅膀就够了。他们去尝试了，但不久后就意识到飞行不仅仅是让身体升到空中。唱歌也是这样，它需要的不只是控制呼吸和声带，还有更多的方面。"

"Mozart was a great mathematician. The rhythms of his compositions affect people's brainwaves, helping children to study. They do better at algebra and geometry while enjoying classical music."

"Amazing! You know, for a long time people thought that all they needed to fly were feathers and wings. They tried but soon realised that it takes more than that to lift a body into the air. It is the same with singing. More is needed than only having control over your breath and vocal cords."

... helping children to study

你们甚至启发了弗里达

You even inspired Frida

"没错，人类需要受到启发才能创造出这些带来如此多快乐的奇妙声音。"

"你们甚至启发了弗里达。我知道她的丈夫被她美妙的声音打动，所以亲切地称她为'弗里达·福格尔桑'。'福格尔桑'在德语中意思是'鸟儿的歌'。"

"That's right. One needs to be inspired to create these marvellous sounds that give so much pleasure. "

"You even inspired Frida. I know her husband was so inspired by her wonderful voice that he lovingly called her 'Frida Vogelsang'. Vogelsang is German for 'bird song'."

"弗里达？你是指来自流行组合阿巴合唱团的安妮·弗瑞德吗？她和三位瑞典艺术家朋友让世界上许多国家的成千上万人与他们一起歌唱。她极其欣赏我们燕子，作为一位姑娘，她连续多日观察我们幼鸟的成长。"

"但这些幼鸟并没产生任何音乐，是吧？他们只是大声叫唤来索要食物。"

"By Frida do you mean Anni-Frid from the pop group ABBA? She and her three Swedish co-artist friends made millions, even billions, of people sing along with them, in many countries around the world. She simply adores us swallows and as a girl spent days watching our chicks grow up."

"But those chicks did not make any music, did they? They were just squealing and squawking for food."

来自流行组合阿巴合唱团的安妮·弗瑞德

Anni-Frid from the pop group ABBA

他们将很快就学会像他们的父母那样歌唱

They will soon learn to sing like their parents

18

"嗯，我们鸟儿的首要任务是找到一位伴侣，产卵并孵化，这是不是生命的奇迹呢？然后，喂养我们的幼鸟，看他们长大——从光溜溜的留巢雏鸟成为有毛的离巢雏鸟，不久后满世界高速飞行，并捕食昆虫。他们很快就学会像他们的父母与祖父母那样歌唱。"

"哦，绝对是这样的！难怪安妮·弗瑞德看到你们燕子时，就想用她华丽的嗓音全身心地歌唱。"

"Well, isn't the wonder of life for us birds first and foremost to find a mate, lay eggs and hatch them? And then to feed our chicks and watch them grow. Watch our little featherless nestlings become fledglings, soon to fly around the world, at high speed, catching insects in flight. They will soon learn to sing like their parents and grandparents."

"Oh, absolutely. No wonder that Anni-Frid felt like singing with all her heart, using her magnificent voice, when she enjoyed the sight of you all."

"我很荣幸! 但你知道吗? 是我爸爸最先找到这个绝佳场所的, 我现在与我的家人仍一起在这筑巢。他先找到一个好地方, 然后开始歌唱, 并夸耀自己擅长飞行。他通过这种方式吸引我妈妈来到这里, 并与他一起度过余生。"

"是呀! 能有莫扎特和阿巴合唱团的歌曲作为背景音乐, 谁愿意去其他地方生活呢? "

"他们只会想待在家里载歌载舞, 像我们在成长过程中所做的一样! "

……这仅仅是开始! ……

"I do feel honoured! But did you know that it was my dad who first picked this excellent place, where I am also nesting now with my family? He first picked a good spot and then started singing and showing off how good he was at flying. He did this to attract my mom to come and spend the rest of her life here with him."

"Well, with Mozart and ABBA playing in the background, no one will ever want to live anywhere else."

"They will just want to stay home and sing and dance like we did growing up!"

… AND IT HAS ONLY JUST BEGUN!…

...... 这仅仅是开始！......

...AND IT HAS ONLY JUST BEGUN!...

Did You Know?

你知道吗?

Adult swallows and their chicks learn to recognise each other's voices. The chicks use begging calls to ask for more food from their parents. Chicks are fed up to 8 times an hour.

成年燕子以及它们的幼鸟学会了辨识各自的叫声。幼鸟用求食声来向父母索要更多的食物。幼鸟每小时需要被喂食8次。

Swallow nests are constructed from about 1 000 mud pellets made from clay, silt and salts. The salts keep bacteria under control. The insides of nests are lined with grasses and feathers for comfort and warmth.

燕巢是用大约1000粒由黏土、泥沙和盐制作的泥丸筑成的,其中盐用来控制细菌。为了舒适和保暖,巢内还有干草和羽毛。

*S*wallows are unique birds in that they mate while in the air. They return to the same nests year after year, repairing old and weakened nests. Offspring build their nests near those of their parents.

燕子在空中飞行时交配，这在鸟类中是很独特的。它们年复一年地回到同一个巢，并会修复残破的巢穴。后代会选择在父母亲的巢穴附近筑巢。

A swallow will bring up to four hundred meals a day to the nest. These meals consist of about 20 insects (compacted into an easy to eat meal), controlling the insect population.

一只燕子每天将携带食物回巢多达 400 次。它们的食物包括大约 20 种昆虫（成鸟要将这些昆虫压缩成幼鸟容易进食的状态），因而燕子能控制昆虫的数量。

Wolfgang Amadeus Mozart for years enjoyed the company of a pet starling. Some parts of the third movement of his 17th piano concerto, KV 453, were inspired by this bird's calls.

沃尔夫冈·阿玛多伊斯·莫扎特多年享受着一只宠物紫翅椋鸟的陪伴。他的第十七钢琴协奏曲（KV453）第三乐章的某些部分灵感来自这种鸟类的叫声。

Mozart considered music to be mathematics that you can hear. Listening to music composed by Mozart enhances mental performance in mathematics and is known as the "Mozart Effect".

莫扎特认为音乐是能聆听的数学。聆听莫扎特创作的音乐能增强数学方面的表现，这被称为"莫扎特效应"。

Had Mozart been paid royalties for his compositions, as is standard practice now, he would have been able to buy the entire City of Salzburg, where he was born.

如果按现在的标准向莫扎特支付作曲的版税，他将能买下整个萨尔茨堡市，那是他出生的地方。

ABBA sold 380 million records over 40 years and this number is still growing. ABBA's album Gold – Greatest Hits outsold the Beatles' classic, Sgt Pepper's Lonely Hearts Club Band.

阿巴合唱团在 40 多年里售出了 3.8 亿张唱片，而且这个数字还在增长。他们的专辑《黄金典藏精选》销量超过披头士乐队的经典专辑《佩珀军士的孤独之心俱乐部乐队》。

Think about It

想一想

Are you inspired by bird song?

你受过鸟儿歌声的启发吗？

Does listening to the music of Mozart make you feel happy? And does listening to ABBA make you feel like dancing?

聆听莫扎特的音乐让你感到快乐吗？聆听阿巴合唱团的歌曲，是否让你想跳舞？

When you build a house for yourself, would you like it to be close to that of your parents?

当你为自己建造房子时，是否希望能离父母近一点？

你什么时候会想唱歌？是清晨醒来时、沐浴时、去学校的路上，还是回家的路上？

When do you feel like singing? Early in the morning when you wake up, in the shower, on your way to school, or on your way home?

Let's have a look at birdcalls. How many different birds do you know of that have beautiful or memorable calls? Make a list of your favourite birdcalls, and share it with your friends and family. You can play them some of the many recordings of birdcalls that are available on CD. Do any of these birds live near you, in cages or aviaries, or in the wild? Find out what you need to do to attract wild birds to your garden so that there will be no need to ever catch, sell or cage birds.

让我们了解下鸟类的叫声。你知道有多少种鸟类具有美妙或令人难以忘怀的叫声？列出一张你最喜欢的鸟叫声的清单，并与你的朋友、家人分享。你可以播放一些鸟叫声，在光盘中有许多这样的录音。这些鸟类是否生活在你身边的鸟笼、鸟舍或野外环境中？想想你需要做些什么来吸引野鸟到你的花园中，这样就不用抓捕、出售或笼养它们了。

学科知识
Academic Knowledge

生物学	雄性燕子的叫声为雌性提供了一个判断潜在伴侣生理状况的机会；燕子用唱歌来表达兴奋、求偶、与其他燕子交流，以及当捕食者到来时发出警报；当人类开发了声带并学会控制呼吸后，声音也得到演化；紫翅椋鸟能模仿人类制造的声音，比如电话铃声、轿车喇叭声或莫扎特的音乐。
化 学	亚洲食谱中的燕窝含有氨基酸、碳水化合物、矿物盐、糖蛋白和唾液酸。
物 理	为了确定燕窝营养价值的真实性，需要采用电子显微术、能量色散X射线显微分析等多种物理化学技术。
工程学	家燕筑杯状巢，崖燕筑穹顶巢，这些结构没有来自下方的支撑，但却支撑了成年燕子和4—6枚蛋，以及蛋孵化后要待在巢里超过3周的留巢雏。
经济学	燕窝位居最昂贵的动物产品之列；全球的燕窝贸易每年高达40亿美元；基于版权的收益；娱乐业在世界经济中的作用和重要性。
伦理学	把鸟儿关进笼子来享受它们的歌声，与帮助鸟儿在家附近筑巢让它们自由自在，但我们仍能享受歌声的区别。
历 史	在希腊神话中，学会飞行的伊卡洛斯飞得太靠近太阳，以致把翅膀粘在一起的蜡融化了，他跌回了地面；燕子在罗马时期被用作信使；中国人食用燕窝汤已有1 000多年历史。
地 理	燕子在除南极洲以外的所有大洲生活；欧洲和北美洲的燕子是长距离迁徙的鸟类；尼日利亚的冬季栖息场所能吸引超过100万只燕子；由于栖息地丧失，燕子处于濒危状态；泰国、印度尼西亚和马来西亚是出口燕窝的三个主要生产国。
数 学	基于斐波那契数列的高度结构化的音乐对心理功能和数学学习能力的影响；数拍子、节奏、音阶、音程、节奏型、符号、和声、拍号、泛音、乐音和音名，所有这些都将音乐与数学联系在一起；毕达哥拉斯用数字来表达音符之间的音程，并从几何图形中推导出乐音；声波能被数学方程定量化描述。
生活方式	我们的生活变得如此忙碌，以至我们忘记唱歌跳舞；父母和祖父母给孩子们唱歌，通常能引导他们发展出对音乐的热爱。
社会学	一些种类的燕子在北美洲和南非被称为紫崖燕；作为食虫鸟类，燕子扮演了有益的角色。
心理学	孩子们每次在练习乐器、演奏音乐时所锻炼的视觉和空间技能，可以强化他们心理与身体的连接；在采用抽象的符号描述存在于脑海中和纸上的对象方面，音乐家和数学家相似；音乐能在情感、精神和身体上激发我们，帮我们记住信息和学习经验；音乐能创造一种高度集中的学习状态，此时大量信息能被大脑处理和学习。
系统论	音乐不只是纯粹的艺术，它也与数学有关联，并在许多人的生活中起到重要作用，唤起生活的节奏。

情感智慧

Emotional Intelligence

狈犬

这只狈犬担心会失去聆听鸟类美好歌声的机会。他特别喜欢听燕子唱歌，所以特别关心这一点。他意识到音乐具有启发和激励的作用，但他对燕子将莫扎特称为数学家感到惊奇。然而，这引发了他将工程实际与生活现实进行对比的思考。狈犬欣赏燕子那激发人类积极心态的力量，这种力量激发人类欣赏自然。狈犬坚持认为唱歌与长声尖叫不同，他愉快地接受音乐带来的乐趣。

燕　子

燕子下定决心把音乐作为日常生活的一部分，但也清楚地表明她在生活中的优先事项是哪些。她并不刻意取悦其他伙伴，但她准备全身心地与他们分享乐趣。她谦逊而诚实，表达出对那些卓越表现者的钦佩，希望将荣誉颁发给那些应该获得荣誉的伙伴。燕子对狈犬提出质疑，启发他透过表面现象来思考问题。然后，在她专注于她最重要的事务——养育后代时，分享了一些有关生活和设定优先事项的智慧。燕子怀念她的父母，并对能在一个载歌载舞的快乐家庭氛围下成长满怀感激。

艺术

The Arts

你会吹口哨吗？如果会，就聆听鸟儿的叫声，并模仿它们的歌声，甚至可以与你的朋友们举办一个口哨音乐会。如果不会，则可以用一块木头来制造口哨。从互联网上查找相关指导，用一些简单的工具来帮助你雕刻。

思维拓展
Systems: Making the Connections

在地球上，声音始终是我们生活的一部分。鸟类具有通过美妙的声音来表达喜悦、呼唤伴侣或发出警报的独特能力。鸟类在城市周边筑巢，用其优美的节奏和旋律鼓舞着人们。遗憾的是，一些人将鸟类从天然栖息地中抓走并运往全世界，迫使它们在鸟笼中度过余生。然而，仍有可能在家附近创造吸引鸟类的良好环境，为它们筑巢提供条件，帮助它们在寒冬和暴风雨中存活。音乐和舞蹈一直是人类文化和传统中非常重要的组成部分，甚至比我们的说话能力还重要。然而，由于缺乏足够的时间和精力在家播放音乐或唱歌跳舞，拜金主义的生活方式常常妨碍我们享受音乐和舞蹈的快乐。技术也扮演了重要的角色，让人们只有一种享受音乐的形式，即限于在电子设备上听音乐或看音乐视频。我们意识到音乐和舞蹈对我们思想、情绪和人际关系的影响，它们能激发创造力，增强情感上的联系。此外，还存在"莫扎特效应"：结构化的音乐改善我们的心智表现，特别是在代数、几何以及时空技能方面。在人生旅程中，音乐会陪伴我们。不管我们的肩上或心里有什么负担，每当我们在黎明时分聆听到第一只鸟儿的叫声，正如同我们聆听凯特·斯蒂文斯演唱的歌曲《破晓》那样，我们都能以微笑来开始新的一天。

动手能力
Capacity to Implement

有多少音乐家给你带来过真正的快乐？我们可以列出一些世界著名的音乐家，但你了解在当地演出的艺术家吗？在周围打听一下，你的老师中是否有优秀的音乐家。或者，你有让你享受音乐的家庭成员吗？你的任务是在社区中发现音乐家和歌唱家，也就是那些能让你享受音乐，或给你带来幸福的人。你愿意邀请他们来举行一场演唱会吗？

故事灵感来自

This Fable Is Inspired by

安妮–弗瑞德·罗伊斯（弗里达·林斯塔）
Anni-Frid Reuss (Frida Lyngstad)

安妮－弗瑞德·罗伊斯（弗里达·林斯塔）出生于挪威，后移居瑞典。她在 13 岁时开始表演，并发展出对包括格伦·米勒、杜克·艾灵顿和考特·贝西的所有曲目在内的各种爵士乐的热爱。她于 1963 年组建了自己的乐队，并于 1967 年录制了她的第一首歌曲。后来她加入阿巴合唱团，成为现代史上最成功的音乐组合之一的成员。她坚定地致力于应对环境问题，是瑞典环境保护机构"自然脚步"（the Natural Step）的董事会成员，也是非政府组织"环境艺术家"的主席。她喜欢花时间去瑞士高山和地中海沿岸，帮助健康领域的前沿企业取得成功，并投身于慈善工作。

图书在版编目（CIP）数据

冈特生态童书.第三辑修订版：全36册：汉英对照 /
（比）冈特·鲍利著；（哥伦）凯瑟琳娜·巴赫绘；
何家振等译.—上海：上海远东出版社，2022
书名原文：Gunter's Fables
ISBN 978-7-5476-1850-9

Ⅰ.①冈… Ⅱ.①冈… ②凯… ③何… Ⅲ.①生态环
境-环境保护-儿童读物—汉、英 Ⅳ.①X171.1-49

中国版本图书馆CIP数据核字（2022）第163904号
著作权合同登记号图字09-2022-0637号

策　　划　张　蓉
责任编辑　祁东城
封面设计　魏　来 李　廉

冈特生态童书

感谢你的音乐

[比]冈特·鲍利　著
[哥伦]凯瑟琳娜·巴赫　绘

唐继荣　译

记得要和身边的小朋友分享环保知识哦！
八喜冰淇淋祝你成为环保小使者！